TRANSITION SECRETS THE T.A.P. OFFICE WON'T SHARE

Your Instructor, LLC.

Author:

Juan C. Martinez, MBA, PMP®, PMI-ACP®

US Air Force Retired

Co-Authors:

Kara Kendall, PMP®, USAF Retired

& Shane Cates, PMP®, USAF Retired

www.YourInstructor.com

PMP® and PMI-ACP® are registered trademarks of the Project Management Institute.

Copyrighted Material

This publication is created from opinions of the author and associated contributors. Nothing in the book shall be considered direction as it is only shared perspectives, opinions, of personnel that have transitioned from the United States Military. The content in this book is copyrighted material of and owned by Your Instructor, LLC. Unauthorized reproduction of this material is strictly prohibited.

Although the author and publisher of this work have made every effort to ensure accuracy and completeness of content entered in this book, we assume no responsibility for errors, inaccuracies, omissions, or inconsistencies included herein. Any similarities of people, places, or organizations are completely unintentional.

Published by Your Instructor, LLC.

ISBN: 978-0-9987060-6-1

All inquiries should be addressed via email to:

info@YourInstructor.com

or by mail to:

Your Instructor, LLC.

1910 Navarre School Road, Suite 6351

Navarre, Florida 32566

10 9 8 7 6 5 4 3 2 1

Contents

Acknowledgement iii

1. Introduction 1
2. Education 9
3. Certifications 30
4. Networking 64
5. Financial Responsibility 79
6. Dealing With The Veteran's Administration 113
7. Job Searching 127

Acknowledgement

Thank you to the many contributors to this book but there are a couple that MUST be singled out.

Many thanks to Kara Kendall and Shane Cates for being the driving force to writing this book. Their contributions with not only providing stories and examples, but also their involvement in co-authoring this book, are truly appreciated. Thank you both for your involvement and I truly appreciate your leadership, direction, and commitment to this project.

To my wife Michelle and Kids Brittany, Ian, Alex, Morgan, and Charlotte, not yet Haro, and now grandkids, Liam, Oliver, and Bay Girl- I love you with all my heart, thanks for supporting me in this mentoring and education journey.

Introduction

At the writing of this book, I have been retired from the Air Force for over 11 years. In this time, I have focused my professional life preparing people/transitioning military members to sit and pass the Project Management Professional (PMP) certification exam.

In my quest to get people certified, I have noticed a trend in people separating or retiring from the military that have little to no plan of action.

This book is my attempt to share experiences not only from myself but from other members willing to share how they could have saved some headaches and being better prepared to serve in their next venture of their life.

From day one of joining the military, every person knows that they are going to have to get out at some point. It doesn't matter what rank you are, whether you are a separating E-4 or an O-4 getting pushed out, eventually everyone must leave the military.

My question to those people is why did you not spend some time in your career preparing for this transition? This book will focus on some skills, techniques, and strategies that will help the students understand what is needed for them to have a smooth transition.

Too many people get out and start to make things up as they go, which lends itself to an inefficient use of their time. Since my post military career has been focused on project management, I will focus this book on teaching students how to manage their life like a project. This should allow students to create goals and measure progress to prepare for the inevitable day of separation or retirement.

The transitioning members should be the project manager of their life and be more goal oriented rather than accidentally discovering levels of unpreparedness.

I spent over 22 years in the military and I have

learned that while you're in the military, the hoops and the hurdles for you to get to the next level are pretty scripted and those members not following that script often find themselves in challenging times because they want to recreate the wheel on their own, instead of capitalizing on the work of the many people that have blazed these trails before them.

It is obviously in the Member's best interest to take advantage of people's lessons that have come before them to allow them to not waste time and to maximize their potential in reaching their next goal.

In this book, we will look at six major areas that each member needs to be prepared in before separating or retiring. Those main areas include education, certification, networking, financial responsibility, requesting Veterans Administration benefits, and how to job search.

It is my personal belief that these six areas cannot be ignored or left to someone else to figure out for you because not having a plan on how to deal with each one of these areas will lead you to be unprepared, waste time, and create unnecessary stress that could have been planned years before your separation.

I hope you enjoy this collection of experiences and take my suggestions as that, only suggestions, as it is your life to live, then it's your decision to charter a path that works for you and your family.

Please remember that I'm not doing this alone, I have surveyed thousands of retired and separated military members and collected my most favorite stories from their experiences to share with you in an attempt to get you to save time and create a plan that works for you depending on your circumstances.

If during the reading of this book you decide that you need help, direction, or guidance feel free to reach out to me on LinkedIn, @Juan Martinez PMP, and I will point you in a direction to assist you.

To start, here is our first recommendation and experience of transitioning from Tracey, a previous service member who retired from the US Air Force.

Tracey S., US Air Force Retired, 20 years of service

"My transition would have been more productive if

I would have started preparing at least 5 years prior to retirement. When retiring it's so many pieces to the puzzle that you have to be aware of and worry about, if you wait until you have less than a year out you are already behind. The current DoD TAPS program only meets the minimum requirements but there is so much information that is left out. The one-week program gives you complete information overload and give you anxiety over your pending transition. I found myself wondering why the transition assistance program is only a week-long for such a life changing situation. Juan, the information in your book is needed because there is a definite need in this area."

As you can see by Tracey's experience and recommendation, start early in your plans to maximize the 6 areas covered in this book…manage your transition like a project and prepare yourself!

I am very excited for the privilege to share mine and other members stories like Tracey's with the journey of the military to civilian transition throughout this book. The goal is to make things just a bit more stress-free for you, also for you to bring this information to those you mentor as well to help them out.

Again, please keep in mind that these experiences and tips still take effort on your part. It's not an easy/hassle free road by any means.

Hopefully you will be better prepared for what's to come, help you ease your mind and give you the necessary knowledge to achieve excellence with your transition process.

Education

In today's military, it is very unusual to find someone without at least a high school diploma. Please remember that is the entry point for adulthood, not the entry point for being a professional.

The expectation if you're going to become a professional, is obtain an education commensurate with your experience, which should eventually lead to a higher paying job. Before you say that money is not everything and that money is not that important to you let's just acknowledge that money is an important key to the world, we live in.

The money is not the most important piece, but rather the opportunities that the money provides, which will allow you to share different experiences with your family. And isn't that what life is about?

If you served in the United States Air Force, obtaining higher education is put at your fingertips through the use of the Community College of the Air Force. This institution provides associates degrees based on experience and continuing education and allows members to obtain an associate degree while working and taking the minimal number of classes.

I have a Community College of the Air Force degree in Instructional Methodologies. While I am very thankful for the opportunity to have moved upwards with my education. I recognize that an associate degree is not enough to get me the six-figure job that I wanted to get after I retired from the military.

At about my 10-year point in the military, I completed my bachelor's degree in Human Resource. Management. Much to my surprise, the Air provided greater benefits to me because of the bachelor's degree because I was selected to attend Officer's Training School.

As a result, this opportunity would have never opened its doors had I not finished the bachelor's degree, no matter who I knew or what I did. Industry also places a special emphasis on members with a bachelor's degrees.

Let's take a look from Armando' experiences and how he viewed pursuing education while in as well as its impact on himself and how he thinks it plays a significant role for a successful transition with other individuals.

Armando C., U.S. Airforce retired 20+ years of service

"A few things I took away when I was in that helped me be successful in my transition was that I pursued my education early. I wanted to have a plan so, I had the choice to stay in and not feel forced due to the safety blanket of military.

Without having a degree and transitioning will only make it more difficult in the civilian sector, so why not ease that worry a bit and plan early and get it done.

The military always makes it a point to over inflate individuals, and I know for my self it was a really hard pill to swallow, knowing that I may not be the greatest as I was led to believe. And here is why I believe this occurs… the EPR/OPR system, awards, and decorations, I think that they do the average military person a disservice and in turn over inflates egos and expectations.

For instance, your supervisor asks you to write a bunch of lines on paper telling them how great you are then, he or she takes its and makes you sound even better. I think this allows individuals to suffer from what I call self-bloat and when you get out and this doesn't happen and it's a shot right to the gut. When you get out employers are not looking at your EPR's nor do they care, what they care about is your education, certifications, and your experience.

Once you decide to leave the military, most of your military accomplishments that you took so much pride in are now not so important to the company looking to hire you.

You have done so much in the past 20+ years but during an interview that twenty years will most likely be wrapped up into a 30 second discussion. If the individual interviewing you was prior military, you may get a five-minute discussion and it most likely will not have a lot to do with you being hired.

I work for myself now and when I am hiring, I am looking at three main points: Skills, Aptitude and Attitude. How you are educated plays a part in all three of those points.

If your degree is in a relevant discipline, boom, skill. Do you prove to me you can you be taught and strive within the *discipline we work with, boom, aptitude. If I know that you have dedicated the time to something to better yourself and reach out for an extension of knowledge, boom, attitude.*

Education plays an enormous role so, start planning early and don't let an organization or other people make choices for you and your future. Get your degree(s) while you're in and can take advantage of TA.

I want to leave you with these important points from my experience that may help. Education is very important, and you should be taking advantage of the benefits the military gives you to pursue a degree. Be a good person and even better at your job, this will create your network for you without you really having to do anything special. Your military experience may be relevant to you but not to everyone else so don't just bank on it and not have a back-up plan. Last thing is, always be looking for the next best thing and remember loyalty is not the same in the civilian world as it is in the military. In my opinion those points can help you have a successful transition from the military."

As a now business owner Armando discusses the importance of education for his hiring process and how it shows much more than a just a piece of paper. It can show work ethic and mental fortitude to a company. He hit the nail on the head about taking advantage of the benefits that the military offers you while you are in.

They give you the funding and opportunity for your education so why not take advantage of them. His points on how much value we place on our military careers and the reality of them just being a small snapshot to employers is not only humbling but should be an eye opener on just how important continuing our education really is. Paired with your military experience, a degree is a complete game changer as a tool for you to have a successful career after the military.

According to indeed.com, a person with a bachelor's degree earns 29% more than a person with an associate degree, and a person with a master's degree earns 41% more than a person with an associate degree. Let me ask you a question, if you have to work 40 to 60 hours, wouldn't you rather make 41% more for taking some classes that you could have completed while you were in the military?

In other words, if you must work for someone else, don't you want to work as few hours and make as much money as you can in that same window? So, my suggestion to you is to take one class per semester as soon as reasonably feasible, and don't stop until you have completed at least a bachelor's degree but keep your eye out towards the master's degree.

Another way to finish your degrees faster is to take advantage of the CLEPS and DANTES free testing out of college classes options. I have never been a good test taker, but I will say that I maximized my use of these two tools because every test I could pass resulted in less physical courses that I had to complete. I found myself going in to take these exams without any preparation, which resulted in failing some exams.

To clarify this point, I took 21 CLEPS and DANTES and only passed 9. But those nine passes resulted in me not having to take 9 college classes, which shrunk my timeline to complete my bachelor's degree dramatically. Since these tests are free to active-duty members, or at least were free while I was in, it makes complete sense to try to take as many as you are allowed towards your degrees to save you time.

For those who might say I'd rather take the class and learn, I applaud your efforts, but frown on your lack of paying attention to the goal. If the goal is to obtain your degree, you can always read books later to make yourself smarter. Don't waste time.

Every now and then I meet a separated or retired member that says that they don't want to pursue higher education. An excuse I often hear is that a degree does not make you any smarter.

To those people, I simply say that is an excuse for not having to work hard. I agree with them, that a degree doesn't make you any smarter. I've met many dummies that have that have degrees, but it does show a commitment to finish something that has been started. Which translates into opportunities when complete.

What I usually tell the people who don't want to pursue their degree because it doesn't make them any smarter is that they are taking the choice out of being smarter by not attending school. The reality is when you need a degree, it is too late to get one because you know it takes years to complete one.

I would rather have an argument with you after you get the degree rather than before you have finished it. Besides, the degree is completely paid for by the Department of Defense, so it seems that it is a better choice to get it done and sit on it and never use it than it is to never get it done and need it for future employment.

There is never a good time to finish your educational goals, it's just a matter of making it fit into time frame you have. If you need to take a semester off here and there, then do it. If you find yourself in a situation where you can multiply your classes by taking CLEPS or DANTES so you take one live class and take two tests, which is the equivalent of taken three courses that semester.

But there is never a good time to sit on it and not do anything about it because in my opinion, not having at least your bachelor's degree will come back to haunt you later in life.

Another idea I hear on occasion is that members want to separate or retire from the military before they pursue their education while receiving their housing allowance. The problem I have with this situation is a twofold issue, the number one reason that I have a problem with this situation is because it'll take the member a couple of years to finish their degree and at the end of that two-year completion date, they would have burned all their networks and their contacts that they knew while they were in the military.

The people they know today will have moved and changed bases by the time they are ready to show off their new degree, and those people will not be able to help them or sponsor them into a new job.

The second reason I don't like this strategy is because their housing allowance in today's dollars is only around $1,500 if you're taking a full time course load, which means that they're going to be receiving $1,500 while attending school, which leaves them very little time to work and earn the rest of the money they need to pay their bills, because I believe you can't make your bills with only $1500 a month.

What that means to me is that they are willing to receive $1500 a month as their housing, but because they did not get that bachelor's degree while they are in, they are forgoing a potential six figure salary when they separate for that $1,500. In other words, I'll keep my $1,500 and throw away the $6,000 or $7000 a month that I could have received had I finished my degree while I was still in the service.

I think the bottom line with this strategy is that it's another excuse for people to not pursue their degrees. What they're really telling me, as a business owner and an employer is that they are going to continue to defer this very important decision until they absolutely must do it and then will work on it when forced, which sends a really strong message of a lack of initiative.

So quit delaying, take one class a semester, take your CLEPS and DATES in between and get it done before you need it. You will thank me by the time you're done.

So… it doesn't feel like I left you much option to pursue something less than a bachelor's degree. If that's the message that you receive from the previous paragraphs, then my job has been done.

Now the next question is then what degree program should I pursue? The answer to that question is one of many options. The first option is to pursue a bachelor's degree in the easiest disciplines that you can accomplish.

While I would never recommend something like Zoology or French literature, a generic business management or organizational behavior bachelor's degree seems to work in many instances for putting a check mark in the box.

As a matter of fact, many colleges provide course credit for experience towards the degrees that align with your current employment. In other words, some universities might give you college credit for experience in leadership, management, or organizational behavior because of your experience while you were in the military but will not give you a single credit hour if you're pursuing something outside, such as engineering or cyber security.

So, the first step is to close the gap with data. I recommend you set up an immediate appointment with your education office to identify what degree programs are the easiest for you to pursue that are still beneficial for your transition out of the military. You might find that you only need to take a years' worth, or two years' worth of college classes to finish a degree.

For the readers that want to pursue a specialized degree, such as engineering or cyber security, then your work will likely take a little bit longer, but you might as well work harder to obtain the degree that you want so it aligns up with your transition.

Stated in other words, if you know you have to take three years' worth of classes in the degree program you want to pursue, you should start three years before your anticipated departure date. I would like to clarify that I am not suggesting you simply take the easiest degree program you could find.

What I am saying is that it is important to separate from the military with at least a bachelor's degree, preferably the one that you want, but if you don't really know what you want, a generic bachelor's degree will suffice for the time being and you can always work on a different specialization as you progress to a master's degree.

What about degrees with specializations or emphasis in areas? My recommendation is to weigh the pros and cons on whether you need to attend more classes to attain those specializations.

For me, I discovered that I would have needed to take three more semesters to obtain a specialization on top of my master's in business administration. After speaking with my college counselor, I was informed that I could get a generic master's degree in business administration without an emphasis and save myself about six months' worth of headaches and schoolwork.

And that is exactly what I did. I finished my degree six months earlier with the generic master's degree in business administration and do not have an emphasis in anything other than just the generic MBA. The funny thing is no one has ever asked me for an emphasis once I told them I have an MBA.

I hope you can see the importance of obtaining your degree prior to getting out. The doors that it will open is accessible no other way. So go out kick those doors open, get that degree, prove to yourself you can do it. The only one holding you back is yourself, there will never be the perfect time to do it so why not now?

Certifications

Obtaining your certification in a professional discipline is one of the fastest and most efficient ways to getting recognized by industry partners. Getting certified in your professional discipline shows your commitment to the to the professional discipline along with your desire to stay current with the changing principles in that area of expertise.

Industry places a special emphasis on people that have obtained their certifications because it shows their commitment to the discipline, their desire is to stay current in that area, and that they are not a part timer doing this out of the garage in the evenings. The real question is, which certification or certifications should you pursue?

The first recommendation I'll make is not to overdo it. Don't go and get certified and everything that you can do because some of them are easy and not recognized by anyone who will ever offer you a job. I recommend you focus on certifications that are going to be within the discipline that you're going to be seeking employment.

For example, for aircraft maintenance people I would recommend obtaining their Airframe and Powerplant (A&P) Certification. For personnel seeking human resources, perhaps a professional human resource certification would do. If you are going to get into information technology perhaps a cyber security or a Scrum master certification maybe perfect for what you're trying to do.

But as of the writing of this book the number one certification that industry is looking for people that want to manage is going to be the Project Management Professional (PMP) certification. It is my recommendation, and yes it might be biased because I do teach the PMP certification, that unless you want to go back to turning wrenches or getting your hands dirty, you should get two certifications, one showing your technical expertise such as cybersecurity or A&P, and a PMP certification because it keeps you in the front office rather than turning wrenches out on the field.

I've had a lot of people ask me if certifications are valuable when they get out and I lump those people in the same category as the same lazy people who don't want to finish their bachelor's degree, they're just looking for excuses to not work hard today.

They'd rather defer any headaches until later and hope that they never have to cross that bridge. But., you will be asked to get certified at some point, so why wouldn't you do it while you're still in the military and Uncle Sam will flip the bill. As of the writing of this book, both the Air Force and the Army have programs to pay for both the certification exam and the prep course, which means members can get a six-figure certification and not pay a single dime out of their pocket.

But the Navy is catching up, the Navy will pay

for the exam will not pay for the exam prep, and I find it kind of funny that students come to me and say I don't want to pay the $2,000 for the class because that's just too much money for them. I believe certifications are so important that they should budget for the $2,000 for the class, and that's not just because I own a class or a business that teaches exam prep courses.

I believe that people should stop drinking lattes and playing golf for a couple of months to save up to $2,000 to pay for the course because once they're certified, then that certification will open doors they never even knew existed.

My company is responsible for certifying thousands of people in project management, agile, human resources, and more. And I see it week after week after week, after people are certified and they upload their resume and their profile on LinkedIn, headhunters come seeking them for opportunities that were not open to them had they not receive this certification.

In many instances I can get a person certified in at least one certification in a month or two without breaking their back, but there is no way I can get someone a bachelor's degree in a month or two because that's just physically not possible. So, then the question really isn't if you're going to get a certification, it's going to be when are you going to get a certification?

Which means that you must plan to obtain this milestone and somehow make it fit into your already busy schedule. So just like in this section, with obtaining your bachelor's degree, you've got to plan schedule it in and focus on it so you can get it done and get it behind you because once you're certified, you're certified for life, as long as you keep up with the maintenance costs and education requirements of the certification.

In most instances, the maintenance costs are a couple hundred dollars every few years and submission that you've done some sort of professional training that will keep your current in the discipline. I am professionally certified in project management and agile and to keep my educational requirements compliant by listening to audiobooks and I claim the hours that it says on the audiobook for the reading of the book.

For example, I recently read Elon Musk book on how he started Tesla, which is a business book. The audiobook said it was six hours long, so I claimed that book under my PMP certification and they gave me credit for the entire six hours.

So, what about homegrown internal certifications, that your organization may sponsor? For example, the Air Force has internal Greenbelt training, resiliency training, and even its own internal Department of Defense Project Management certification training.

My recommendation is while you are serving to go ahead and take all those trainings because you must be compliant with them while you're doing your current job. Furthermore, if you decide to get out of the military and come back in and work for that same branch of service those certifications will be recognized by the same service.

However, if you decide to pursue a career outside of that service or jump to a different service you will find yourself justifying or explaining what those certifications are important in your new job. Civilian certifications don't work like that, they transfer immediately to the military.

For example, if you have your Department of Defense certification in project management, that is only good if you become an active duty or civil servant project manager in the military, not as a contractor.

But if you have your Project Management Professional (PMP) certification, that will work for both as a contractor and if you work for the government. Remember that when you seek certification opportunities that the certification should be few.

They should be focused, and they should help tell a story on your resume. They should help emphasize that you're a leader, that you're a technical expert, and that you should be trusted and relied on in that area of discipline.

People that have multiple certifications in all types of disciplines are only sending the message that they are professional student and are not focusing their efforts on their current job. I'm not saying not to get multiple certifications in many areas. All that I'm saying is that your current resume you are sending forward for a specific job employment should only list the ones that are required to tell that story of being a technical expert, professional in the area, and being trustworthy.

In this revision of your resume, you might find

yourself potentially eliminating all your non relevant certifications, or at least downplaying them later the bottom of the resume.

So why should I get certified? There are many reasons why you should seek certifications, but here are a few. Certifications makes you more credible, increases your marketability, your boss likes it, and increases your chances for promotions.

They make you gain more professional job satisfaction. Furthermore, the help you earn more money and get new skills or polish up existing skills in your job. And later, when you're finally ready to do your own thing, if that's where you want to go, you could become a consultant because becoming a consultant without being certified, is like having a boat without a paddle or a motor.

Let's talk about each one of these individually, and I'll explain why they're so important.

Having your certification allows you to gain credibility in your workforce. It shows people that you're committed to the discipline that you're serious and that you do want to move forward in this area of expertise. In today's, leave your job fast and get another one world, obtaining your certification in that discipline shows the hiring officials and your leadership that you're here to stay for a while.

After all, why seek a professional certification if you know you're going to get a different job in a different area? So, the certifications bring with it a certain amount of credibility towards your name and to show commitment in the area.

If it's one thing I've learned over the last 30 years of working in the government and in industry, is that credibility is everything. Your name is the only one you have, and when you tarnish it It's hard to come back. People know who you are, therefore, this credibility adds strength behind your name and almost instant trust within that area of expertise.

Once you have the expertise you've increased your marketability in that workforce. One of the common themes I hear when people get out of the military is the word loyalty. I have loyalty just like everybody else, but it's been changed a little bit.

When I was in the military, loyalty meant working towards your organization and blindly accepting the challenges that came forward because that was something that the Air Force called service above self.

While I was in, that is a terrific tool to keep me focused and ensured that I kept the mission of the organization first. But loyalty when you get out of the military, is not the same. There is a reason why you must complete a full pay period and wait a week before you get paid.

It's because industry wants to sure they calculate exactly how much money they owe you, so they never owe you any more money than just one pay paycheck. Since industry doesn't have as much loyalty towards you as the government does, in other words, they will gladly replace you with someone faster, cheaper, leaner, meaner if they could.

That means you must maintain your marketability and one of the fastest ways to stay relevant in your discipline is to get and stay certified while maintaining your networks. Getting certified in the areas of expertise means that you can remain marketable in that area so you could pivot if either you are replaced, or you want to seek other opportunities because remember, loyalty is not the same while in the military as it is off base or post.

The only expectation is that you at least give him a warning of a week or two before you separate the organization, unlike the four months to a year before you must wait to separate from the military. Increasing your marketability in the job means that you can stay nimble, and you can move opportunities arise.

When these opportunities pop up, it is because an employer has a preference towards something that you have that they cannot or do not have from their current team. Employer preference means that they want you to have the certification even if you don't believe that it's important to you.

I've had many people tell me that having a certification is not important in their eyes. But since I'm the one who's already retired, I can tell them that it's not what they think. It may correct in their eyes, but if it's what the employer prefers for them to have you then it becomes important and a necessary requirement.

Members seeking six-figure jobs need to be aware that there is likely a stack of candidates and it is their job to put what is referred to as "legal discriminators" in their resume to make sure, or help, their resume float to the top.

Examples of legal discriminators are education, experience, certifications, network connections, etc. Your job is to make sure that your resume has as many of those legal discriminators in it to make it difficult for your employer to not offer you an opportunity.

So, it's time to focus, the certification is not for you, it's for the employer. Do your research in your research and find out what certifications are important in your career field. If you don't know, simply reach out to me and I will gladly mentor you to ensure that you are pointed in the correct direction.

Notice how I said it's employer preference and not your preference. If you want to get certified in basket weaving, that may be fun, but it may not be what the employers are preferring at that time.

As you're working on obtaining your certification(s), please remember that when you are finally finished getting certified, it does bring a great sense of pride, almost like a big weight has been lifted off of your shoulders and there are several reasons for this.

One of them is because you are now not indebted to the government for the money you borrowed to get this certification. In other words, you got certified. And they are no longer going to threaten to take money out of your paycheck because you haven't finished.

Let's face it, it does feel good when you close that government requirement. But what about the personal requirement of getting it done? How does it feel when you're finished with the certification that you been working for a couple of months, and you finally get it?

Many students have told me that it feels as if they're lighter than a than a feather. It feels like they have taken a big weight off their shoulders and now they could focus on other higher priority things.

Yes, it's true that getting certified could be like

putting a check mark in the box into many people, and some hate that game. Please remember many of the transition steps seem like a game but most are mandatory for a smooth transition. I recommend you find a certification that fits within your area of expertise, focus, plan, and work hard for it, and then celebrate the accomplishment when you're finished.

I often tell students, when you get your professional certification to take their family out to dinner and not to a fast-food place or go through a drive thru, but a good dinner and celebrate your accomplishment. Getting certified brings a huge sense of personal satisfaction because it is the receipt of all the hard work that you've been putting into this certification.

Another reason why it feels personally gratifying to obtain a certification is because it validates all the hard work that you've been doing over the last several years into a piece of paper that shows that you're certified in that area, which means, the other certifying body agrees with you that you are smart and capable in the area as you said you are.

After all, any fool can go to the store and buy a paintbrush and call themselves a painter, but only the person who went to the actual painting school can call themselves a certified painter. Let me ask you a question, who do you think is going to get that big painting gig? A certified painter or the guy that just went to the store and bought a paint brush? My guess is the certified painter.

One of the biggest reasons of getting certified is simply to make more money. A certified painter will likely make more money than a non-certified painter. While I was in the military, I was an acquisition program manager and I got to work through government contracts as a daily routine.

I noticed that that the government had restrictions on certain jobs within contracts and required them to have a "senior person" in that position. Ironically, senior person in the position could not be proven with just a resume. Many times, it had to be proven with something other than their time on job such as certifications.

I have seen it over and over and over and over again where a project manager will make $70,000 a year but could make up to $130,000 or $140,000 or even more a year if they were simply certified in project management.

Earning more money is an absolute benefit to being certified, it is my expectation that you will make more money if you are certified. As of the time of this writing of this book an average project manager that is certified in project management through the Project Management Institute (PMI) makes approximately $20,000 more than someone who is not certified.

In other words, if the government is going to pay the full fee for you to take the class and the exam and all you have to do is study and take a test, why wouldn't you want to make $20,000 more a year for a certification that you can finalize in a couple of months.

I understand, like I said in the previous chapters, making more money isn't necessarily important to everyone, but it doesn't mean I'm going to throw it in the trash either. Why not focus your studies on a certification, get certified, get that extra $20,000 and if money is not important to you, then donate it to someone who can use it?

Some of you probably smile reading that line thinking I'm not donating my extra $20,000 to somebody. Well, you're probably not the same person that would say that money isn't that important to you. But if you're telling me that money isn't that important to you? Then why not make more money and donate it where you feel is most impactful as opposed to not taking it from the company, letting them spend it or waste it wherever they think is important.

Like I said before, I think people that don't want to work on their certifications or argue about it or just simply making excuses and trying to defer that requirement as long as they can until they're forced to do it, and by then it's too late.

The opportunity would have been gone and someone else has the job and you're stuck looking at it from the outside in, probably saying, well I guess I didn't need it. I'll wait for the next time and finish it then.

Another major reason why someone would get certified is to gain new skills or new knowledge in that area. As I mentioned before, once you're certified, there will be an education requirement and a financial requirement for you to maintain your certification.

Most certifications don't have a reoccurring test that you have to take once you're certified. In other words, you test once, and you're finished. But there is an educational requirement and a financial requirement that must be made every couple of years. Since I'm certified in project management, I have to pay $150 every three years and submit 60 hours of continuing education to prove that I'm staying current within the workforce.

The $150 is a piece of cake, I write a check or do one bill pay and it's done but figuring out how to get 60 hours of education into my studies within that three year can be a challenge.

Like I mentioned, I do a lot of audiobooks and I always get more hours than I need, or I get my hours way earlier than I need them, so it's not that big of a deal. I can do 10 audiobooks in three years. No problem. But other areas to gain expertise that you could do to satisfy the educational requirement would be to continue on with that master's degree that I've been recommending to you.

Every class you take towards your master's degree can count towards maintaining your certification in most areas. In other words, if you already have to take master's degree courses in a discipline, and you're certified in something you could claim the classes towards your master's degree as your continuing education for your certifications.

Which is twofold, it helps you maintain your certification and allows you to gain new skills and knowledge in the area of discipline that you're pursuing. Another thing you could do to gain knowledge in your area is to watch YouTube videos.

My morning routine includes drinking a cup of coffee and watching at least one YouTube video every morning. I try to watch something that I don't have any expertise in. For example, I might learn how to build a new fishing dock today, or I might learn how to wax your car tomorrow, it doesn't really matter what I'm learning because some of the skills that I learn in those unrelated fields lend itself to some type of project that I may be managing later.

So as a result, I'm always polishing up my skills and knowledge. By watching these YouTube videos to make me more well-rounded expert, obviously, to gain knowledge in my area of expertise, but if you would/should watch videos within your area of expertise.

For example, if you're a cyber security expert, you may want to watch YouTube videos uploaded by cyber security experts in that field, I do that also. It's just in the morning while I'm having my cup of coffee, I try to gain knowledge in areas outside of what I'm dealing with for the rest of the day.

It's my personal belief that when you gain education or knowledge in the area of expertise, they should try to dovetail anything you do, such as reading books or watching videos to help you with your job.

Find a way to link learning to what you're doing in your job and that way you can claim the hours of what you're learning for your job. For example, if I watch your video on how to make a boat dock, which is likely something I'll never do, I can claim that skill on my PMP certification educational requirements by calling it a class in boat dock installation projects.

This allows me to capitalize on learning the information and using the hours towards maintenance of my certification which makes me smarter, and I get to claim the hours for my certification.

And lastly, this may be way out for some of you, but becoming an independent consultant in your area of expertise can be the ultimate reward in becoming a professional.

Most people I meet are completely satisfied and

happy working for other people, for their entire career. It provides a steady paycheck and provides a level of planning and accountability for their check checking account, which provides them that stability that they and their family require.

But for a few of us, that will never suffice. Some of us want to branch out on their own and become wealthy and become independent consultants or independent contractors and get financially recognized as such.

One of my mentors told me many years ago that a professional project manager consultant should make no less than $400,000 a year. When I heard this, I kind of laughed but over the last ten years or so I have found myself into the big ballpark dollar figures because of being an independent consultant. The problem with that is that it also creates big ballpark risks as well.

In other words, when you have a job working for an employer, there are very few risks associated with getting your paychecks. But when you become independent, you have to rely solely on the business you bring in. In an industry like the one I'm in, which is certification exam prep, you will find that once your student has taken your course, there's nothing else to sell. Which means you don't have as many repeat students as, let's say, a restaurant where you're hungry every day.

Therefore, you might go with dry spells, or periods of time without customers, so it can be very nerve wracking if you're not financially stable. I understand that some of you might say, well, if you're making $400,000 a year or more, then you don't need as many customers you can afford to have dry spells.

The problem is that we're Americans, and the money we earn is usually spent immediately. Most people I've met can't hold onto a dollar to save their life. Which means that you really need to pay attention when we get into our financial responsibility section coming up.

Becoming an independent consultant provides you flexibility that you would not normally get if you had a normal 7:30 to 4:30 job, however, for those of you that think that you don't want to work 40 or 50 hours, I would not recommend becoming a professional consultant because you're never really off. You're always checking emails, you're always responding to client requests, you're always providing proposals, you're always upgrading to the next big thing so you can become more marketable. It's a never-ending job, trust me, you will earn every dollar you make.

So whatever reason you choose to get certified, remember don't overdo it. Get certified in the area of interest in order to improve your resume to help bring into focus your commitment to the discipline, being a technical expert, and also a trusted person within the within that job.

Networking

You will never have a bigger network then you do the moment you transition for retirement. Even if you think that you don't, I promise you, you do. Striving to create and nurture those relationships is one of the most valuable steps you can take to become successful.

I have a few crucial steps that will inevitably bring you success in your networking journey. First, you must make sure that you have yourself together. This starts with always making a great first impression. It is extremely important when establishing relationships to have that best foot forward.

Make sure your dress and appearance is always where it should be or above where it should be. Be respectful to everyone, and if you are in uniform, always adhere to the proper customs and courtesies. Be the individual who people want to work with and surround themselves with.

Put in the work, and strive to lift your subordinates up, don't take the persona on of they work for me. Work for them, build them up, make them great and your team will always work harder for you. Be a servant leader, your team is the single most important weapon in your arsenal. Work for them and you will see the investment hit new heights within your organization.

Make it a priority to help others, I am not only talking about your leaders and individuals in your unit, but that should also be a no brainer. I am talking about individuals outside of your office, build those relationships as much as possible and any time you can.

Meeting people outside of the office plays an important role on networking. This can be done through volunteer functions, events, clubs, associations, etc… You should always be striving to help others and not in a selfish way, be genuine and do on to others as you would want someone else to do for you.

Having a good relationship with your leaders and bosses is another great networking tool. I am not talking about an unprofessional relationship, or you have to be drinking buddies on the weekend here.

I am saying work hard for them and make yourself irreplaceable, by doing this will not only prove to them your value but to others as well. You never want to burn bridges; this goes for everyone you come in contact with. You never know if that individual could help you later for a letter of recommendation or possibly, he or she becomes your boss one day.

Find mentors, I am not just talking one, have many. None of us are perfect at every facet of life, success is built on the shoulders of giants. Search out these giants and pick their brains, get to know them and watch how they conduct themselves. You will need help and role models, just like others did before you and well after you will need someone to look up to.

Become a mentor and help others to achieve what you have, being selfish or arrogant will get you nowhere professionally for any length of time. Make it a point for yourself to always pay it forward, no single person can achieve greatness all by themselves. Be the greatest version of yourself, push yourself to be better and to be a person you yourself would look up to.

Being able to keep and maintain relationships is vital in the civilian world for success. You just never know when you might be able to help them out or vice versa. Being able to market yourself with a true helpful mindset will always be helpful as well. Make yourself as approachable as possible to those whether it's to provide mentorship or just to help them out.

An important philosophy that I would consider

is that absolutely no one wants to help someone if they know the act wouldn't be reciprocated. I am still in contact with countless individuals from my military days. I have gone out of my way to help them, and they have done the same for me. So why wouldn't you want to create a network and help others and yourself with our push for success.

It is critical for our network that we always strive to put out our best foot forward when establishing and maintaining our professional relationships. There are a few things to consider when maintaining a working professional network. What do you bring to the table for the relationship to strive? How will you establish contact and work with each other? How are you going to track and stay relevant with your network?

There are many ways to go about this and I will do my best to discuss and layout a map to achieve this to the best of my ability. One thing that is an absolute must, go to an online business card company and create business cards for yourself. It doesn't have to be anything fancy; the whole purpose of these cards is just to get your name out there and give someone something to reference of you.

Like I said nothing fancy and not all of your military information needs to be on it, remember you are working to move away from your rank and current titles. Just your name, email a phone number and a one-line description of what you bring to the table. For instance, here is a quick example.

This will be a great tool for your arsenal to boost networking and marketing of yourself. Keep them in your wallet, vehicle or on your desk at work, when you meet someone and you're finishing up the conversation it so easy to just hand them your business card.

They may take it and throw it in a drawer but if their company or any of their associates ever needs anyone with your skill set, they can always pull it out and just like that you are now relevant in the conversation.

I would encourage you to attend as many meetings relative to your skill set as possible. To use myself as an example, I have my PMP certification and I seek out any and all types of workshops, seminars and meetings that are free and open to the public and attend them. Not only for the passing of knowledge but also for the socializing prior and after. You bet I am going to have a pocketful of my business cards that I can hand out and pick the minds of others successful individuals in project management.

These gatherings are an amazing way to stay relevant in whatever career field you are planning on going into. I would start doing this as soon as possible. If there, is one taking place on your lunch break proudly wear that uniform and become a sponge.

Don't forget to bring your business cards and hand them out. Not only does this get your name out there but it shows your drive and interest to possible future employers. I encourage you to go out and find mentors and individuals to ignite or stoke that spark or fire you have for that next chapter.

When it comes to technology where it is today it creates a new way of doing things. This is beneficial for potential employees and employers. Take a look on Eric's thoughts on this.

Eric T., United States Air Force, 20 Years of service

"Preparation for one's departure from the Armed Forces is vital to a successful transition. Admittedly, it is time consuming, stressful . . . and dreaded but it is extremely important.

I found that starting 24 months in advance made all the difference. During those 24 months, this gives the member time to focus on their resume, finishing education or certifications, networking, preparing for potential interviews, and while it will be too early to apply for jobs, they can at least start looking at their potential market.

There are numerous ways to build your network. One of the fastest growing ways to network is LinkedIn. People can search by country, region, state, city, name, or job title. A simple post introducing yourself, that you're leaving the service, and what your career aspirations can literally be life changing.

In my preparation, I used LinkedIn to get me connected with companies, talent acquisition, and veterans working in the field I was hoping to transition into following my service. Because of that platform, I was able to have numerous calls in which professionals reviewed my resume and provided recommendations as to the content the resume contained."

Eric is spot on with his opinion of the importance

of utilizing technology and these sites to make the most of the "dreaded" transition.

With the technology in today's world networking over the web is endless and if you refuse to make the technological change you are at a huge disadvantage. There are countless social media outlets that are great for networking and professional development. One that I will talk about that is enormous right now is LinkedIn.

If you do not have an account yet I need you to stop whatever you are doing and make one as soon as possible. This is used by almost every hiring search engine for possible job candidates. As far as a professional directory goes there is not one out there that I've seen yet that compares to it.

You have your work experience, resume, certifications, education, and location on it. The site also allows you to put on there if you are looking for work that will put a ring around your picture. This allows companies to headhunt you when they see this.

The days when we used to have to get all dressed up and go business to business, handing out our resumes are gone. That is not saying you won't have to do that but, that interaction will most likely not be the first thing the hiring manager has seen on you when you walk in the door.

Think of it as, every call back or interview you take will more than likely be the second or third interview. This being said, your foot is in the door and these sites have done most of the work for you already.

Let's take a look at this suggestion…

Chris C., US Air Force, 20 years of service

"I retired as a K-9 handler after 20 years of service and found no good paying handling job locally. I did find a handler job one town over but paid less than an E-5 pay and I didn't think the risk was worth the pay so I reached out to my friends. Turns out, an old friend of mine worked on a government contract working with computer security and despite my lack of knowledge on computers, I was brought on for my physical security expertise.

I recognize this job was not a perfect fit for me but it met y needs until a job that I really wanted opened up. I suggest you keep a good network of friends that may help you out that can vouch for your character as you will never know where one job may lead."

As you can see, Chris made the most of his opportunities and with a little patience, he was able to maneuver to a job that better suited him without losing income. Remember when you are getting ready to leave the military, your network will never be where it is at that moment. Nurture those relationships and build upon those relationships solidifying that bond. Nobody can do this alone, help someone out that needs help and at some point, you might need help to get things moving.

Financial Responsibility

WARNING LABEL: Before we begin this chapter, please remember that I am not a financial counselor, or a financial advisor and the ideas presented in this chapter are steps that worked for me or are a collection of ideas that worked for other people and not to be taken as direction or guidance in your financial future. Adult supervision is advised.

Managing your finances while you're in the military is kind of routine because you know the amount of money that is coming every other week with an almost 100% certainty that it will arrive. As a result, you're allowed to make mistakes that could be fixed within a day, or two, or even a couple of years if you got yourself into a bad financial situation.

When you get out of the military, you lose that financial security and having a secure paycheck every couple of weeks goes away and your paycheck is only as good as the number of hours you have logged into your timecard. In other words, you have to work for your money. I don't mean that in some sort of negative or insulting way, but the reality is if there is no such thing as an automatic paycheck when you get out, and as a result, you're going to be forced to manage the money you have in an effective manner to spread across all of your bills and financial responsibilities for today, tomorrow, and hopefully for the rest of your life.

After all, the point of having a job is it to never have to work again. The only reason I have a job today is because I don't have enough money to stop working, so I continue to work, save money, and make the money grow and do the things that I want to do.

One of the biggest financial lessons I learned while in the military wise to live within my means. When I was an E-3 I lived within my means… yes, I got wrapped in a car payment, and yes, I bought things with credit, but my bills never exceeded the amount of money coming in and I was able to dig out of those financial situations.

As I matured financially, I really didn't need more money to live, I just wanted more money to experience things that I believe would make me happy. For example, the day I bought a brand-new car the actual purchase did not make me any better of a human, but it did provide me the intrinsic feeling of driving a new car which provided its own set of mental benefit.

As many of you know, those mental benefits go away the first time somebody dumps a juice bottle in your car or does something in the backseat making your car no longer have that new car smell.

Generally speaking, a lot of people find some level of balance where they just don't need the extra money to live so the extra money could be used for some benefit instead of to buy more toys that only provide a temporary 'high." Some people I've met act like the retirement or separation date is the day they're going to start being smart with their money, and let's face it, if you have 20 years of bad financial habits the chances of you fixing them simply because you're separating is probably slim to none.

I recommend when you separate or retire from the military they should be as debt-free as possible, in other words get out of your car payment, get out of your credit card payments, get out from underneath your student loans if you have any.

The only payments that you would have, are those that you need to live in such as potentially a house payment, and electrical, water, sewer, but no car payments, no cable TV payments, or credit card bills.

One exciting aspect of retirement is getting

that well deserved retirement check. People get really excited about the amount of their paycheck from their retirement and there's a couple of things they may not be aware of.

The first thing is at the end of the year you are going to be taxed federally on your retirement money that you made from the government, in other words if you're making $2500 in retirement as the retirement calculator might say, you're not going to actually get to keep that much. You're going to have to give some back to Uncle Sam and possibly some states taxes, and your local township may also get a cut! When I lived in Ohio the township charged me a small percentage as well so by the time I got my retirement check, it was cut three different times before I saw one cent.

So, what can you do about this? Please remember I'm not a financial counselor, but perhaps you can choose a state without state taxes? If that is not an option you should live within the money that you actually going to get, not what is expected. Don't plan on the $2,500 because you don't really know what you're actually going to get. So how you know exactly what you're going to get?

The easiest way is to find somebody who's retired where you are retiring and has the same rank and time in grade as you. This is the most reliable method to provide you some really good estimates of how much money are going to get.

Notice how I recommend that technique over using a government provided calculator. There's generally nothing wrong with the government provided calculators on the websites, the problem is that you don't really believe them until you see it in your actual paycheck or that of a trusted mentor.

Another thing to consider when you're thinking about your retirement check besides taxes, is how many other responsibilities you may have that are going to deduct from the amount of money provided.

For example, child support, alimony, tri-care bill, dental, life insurance, etc. My current Tricare bill, as of the writing of this book is about $480 a year which is deducted immediately from my retirement check before I get it another deduction. My dental bill is approximately $70 a month, and lastly, I have some life insurance policies that come directly out of my retirement check.

Since I don't want to have a lapse in paying my medical, dental, and life insurance I have them immediately taken out of my retirement check.

A major point of contention with some people is paying off their cars or house. Some people believe that they will always have a car and house payment, and to those people I say yes! You will, but financially secure people won't. I recommend as soon as possible start making extra payments to your house and car and roll over any equity in any houses sold during your time in the military into your final retirement house.

I see many people that make great sums of

money in selling their homes while in the service only to trade that equity into toys and bigger cars, some with payments. I know it seems like an impossible to some, but you can do it. It may take 20 years from now but that is better than 30 or more years.

If not now, when is the right time to stop giving money away to some lender to let you borrow their money? Please remember that by the time you pay back the loan following the entire term f for a home loan, you will likely pay them back, in some instances, more than double what you borrowed!

Some of you may think there's no way that you're ever going to pay back more than double of the money that you've ever borrowed and I'm telling you right now that you are absolutely 100% wrong! When you purchase a home if you were to finance it for 30 years you paid more than double the amount of the home that you're buying in other words if you buy a $200,000 house at today's interest rate of 5.6% you will pay back more than $413,000 in total cost.

In other words, you paid the back $213,000 of nothing but interest for the pleasure of having the bank open for that loan open for the for 30 years.

If paying off the house is a bridge too far, let's try paying off the car. This task should be easier to do because car payments are naturally done within five, six, or seven years, right? Nope. My experience is that many people I meet will never make the final payment on their car because by the time their final payment is due, they're already trading it in to buy something else.

Let me tell you, I've heard every excuse under the book, and that's exactly what they are excuses, I've heard people say I need a new car because I just couldn't depend on the old one, whether maybe some of that is true, I have not found a car repair bill that is worth having a brand new car payment, in other words if I had to replace my transmission for $4,000 that is still cheaper than buying a new car, and in today's dollars they could be upwards of $60,$70, or $80 thousand dollars or more.

So why would anyone take that kind of responsibility when they know they're separating from the military? Another excuse I've heard is "why am I gonna make money if I'm not gonna spend it"? While it may be true that tomorrow is not guaranteed, the idea of you only live once has taken many people to the poor house.

It is true… you only live once, but that is not an excuse to be financially irresponsible or to begin a financial commitment beyond your level of guaranteed paychecks.

The last topic I'll talk about is the last excuse I've heard people talk about is that I needed a bigger car because my family has grown. I do believe that may be true in some instances, I mean after all if you've got three children putting three car seats in the backseat is probably not gonna work.

But I've seen people do this just to justify their decision, for example I've seen people by trade in a

Honda Civic with driver seat, passenger seat, and three seats across the back for a Toyota Rav4 which is exactly the same seating it's just a taller vehicle.

In other words, the fact that their family grew is not aided by the car they bought because this other vehicle is basically a one for one trade except you have a higher payment. I think in the game of financial responsibility we will always find people that make excuses to the point where they believe them, and it doesn't have to be that way.

My recommendation is a couple years before you get out, whether it's one or two or three years, eliminate those bills so when you separate you have as few as possible. This will allow you the most flexibility to not have to take a job you normally wouldn't take because last thing you want to do is to be desperate because your handcuffed to your bills.

Get those bills down as far down as you can and live within your means as fast as possible and then once you get a real job when you separate from the military, and you know what your paychecks are going to be then get financial commitments that you can live with.

Another recommendation that I have for you is when you start to live within your means, be deliberate with saving for those anticipated reoccurring bills such as vacations or holidays. It seems to me e that most of us don't send money aside for future plans such as vacations or emergencies and rely on our credit cards to bail us out.

On top of not living within our means, retired military members that take on civilian jobs don't always participate in their company's 401(k) plans because in our minds we already have a retirement plan, while that is true, there is a major flaw to it and that is that our retirement plan is only going to grow basically at the rate of inflation for the rest of our lives. Think of it this way, if you retire today making $2500 a month, 20 years from now you're going to be making just a little bit more than that, which means your retirement check won't buy as much as you thought it would in the future.

If you can participate in a company's 401(k), they will likely contribute to your investment which means you can stay ahead of your shrinking retirement check.

I believe the reduction of debt is so important prior to separation that I would even urge you to get a part-time job and see if your significant other will contribute, if not already contributing to the removal or elimination of your debt before you get out. It is not a secret and families suffer because of financial responsibility so why are you introduce new or financial commitments if you don't have to.

I recommend delivering pizzas on the weekends or taking on side hustles in order to separate without debt because side gigs can make a very large on contribution to eliminating the debt. I already know some of your say why would I want to have a side job? I barely have enough time for the one job.

Please consider that at some point you are going

to figure it out I just hope you figure out my reading my reading this book rather than some sort of financial hardship. I really do hope that you learned his lesson sooner than later because these lessons are paid in dollars which restrict your future flexibility.

Let's take the story of my friend John who retired from the military and made one financial decision that he regretted.

John K, US Air Force, 21 years of service

"The biggest financial issue that I created before I retired was, I purchased a brand-new vehicle for my family for months before my final retirement date.

I did not have a job, but I didn't know I was moving back to my family farm in the northeast corner of West

Virginia. Where I inherited a small piece of property, with a small fixer-upper which I knew was going to require a lot of money to fix up to the standards of my wife and I and our family.

Instead of driving the vehicle that I had that was already paid for and met our family needs, I decided to buy a brand-new vehicle which means that my retirement check of $2800 a month was already subtracted for the next six years to the two of about $700 just for the car payment. Not to mention the extra car insurance because now I had to carry full coverage which brought my car payment with insurance up to approximately $800 to $825 a month.

As a result, my retirement check went from $2,800 a month to $2,000. I don't know what came over me to do this, it felt like the right idea that time and part of it was *that I told myself that I have a job now and I qualify for the car payment and if I don't have a job when I separate and we can't get a new car without employment, but there was really nothing wrong with my old car.*

I only did it because I was scared that I wasn't gonna get a job, but fortunately for us I got a terrific job making fantastic money so I could afford a car payment

The truth is that I had that job before I bought a new car, I probably would've bought a different vehicle because my budget would've been more realistic to what I could afford, it would've meant an upgrade but now I'm stuck with a car that I could afford at the time that isn't really what we wanted."

As you can see, John committed himself financially a couple months before his retirement unnecessarily. I want to open your eyes to know that if you can't get a job after you separate why would you get a car before you separate?

Why increase that stress level and I kind of headache months before you separate so when it comes to financial commitment? I recommend you try to eliminate financial distractions and noise before you separate or retire.

As you grow financially responsible, you should have a plan for the money you're going to make because if you don't have a plan, money seems to slip out of our hands and you end up losing it all, and I don't want to see that happened to you.

Let me share with you with what my wife and I do to keep our finances secure. What we do is our quarantine our US Air Force retirement check and our VA disability check by depositing those two checks into a bank account in Texas that we don't even have an ATM card for.

Yes, we do have a checkbook for that, but we don't have the ability to instantly withdraw money from that account because we want to have to think about any withdraws prior to making them. From that account we have automated our bills to allow for a controlled process to pay our bills and the remaining funds stay in the account for savings.

As a result, extra funds accumulate in the bank account and because we don't have a lot of bills it accumulates very, very quickly. Before you ask yourself, or before you ask me how do you do other stuff like fun stuff? What about vacations and life experiences? That is where my post retirement job or the income I make it my post retirement life come in to play.

My post military jobs and investments provide my family and I what we consider to be a financially fulfilling and experience rich life. We are not rich, and we're certainly not famous, but we do live a very comfortable life because we don't have any debt, and the bills that we have are generally must have bills such electric, water, cable, and internet, and the rest of the money is sent in the savings account or investments for emergency or vacations.

What I like about our situation is that I am incentivized personally to work my post retirement job to make extra money to buy those other things that are not required for us to live, which gives me a reason to work.

Let's take a look over at my friend Mary did with her finances when she got out of the Navy.

Mary L, US Navy, 12 years of service

"When I was in the Navy, I learned that money was not something that was going to be given to me. I needed to earn a paycheck to sustain my life I wanted to start a family without the commitment of being abroad.

I took a job that met my needs and discovered that my Veteran's Administration disability was approved for 40%. I was thrilled! This meant that my medical was free at the VA clinics, and they would pay me a tax-free check of about $600 per month. Since I was able to live within my means,

I was able to save the VA payments for about 4 years for a down payment on a home.

I know that I could have used my VA loan to get my house financed with no money down, but I really wanted to make a big dent in the amount of money that I owe the bank so I wouldn't have to pay such large interest. As a result, I was able to save $600 a month for about four years and then I was able to put all that money down, which is approximate $30,000, and the best part is I never needed the money and it was all provided to me by VA which I used towards the VA loan.

In essence using VA money to pay the down payment and I used my VA home loan benefit to purchase the home. I could've bought myself another toy or bought myself more stuff, but I kept my eye on what I really wanted, and that is *to reduce the amount of money that I owed so I can live a free financially free life."*

In my opinion, Mary played her cards very well in that situation she had to financial discipline to save her money focus on a goal and execute that to the tune of saving $30,000 and the associated interest on top of that $30,000 which on a 30-year note could be an additional $30,000 or $40,000.

In other words, she paid $30,000 and saved an additional $30,000. Again. I'm not a financial planner or consultant I'm just giving you some examples here.

So, what is the right financial answer? I can't tell you that, but I can tell you doing what your friends do financially is not the right idea. I believe most Americans don't really manage their money well. Americans believe they are prosperous because they have Harleys, and Jet-Ski's, and F250s and then they are broke.

I believe that is not the measure of financial success, in my opinion my measure of financial success is financial independence. The idea that you can live within the money that you retired with or any veteran's administration, if you're so fortunate to get some, if you can keep your bills underneath that number, that provides you the financial freedom to start your next chapter of your life.

This allows you to accumulate riches and wealth for prosperity rather than to survive so the next time you go and sign your name on a direct line for a new loan.

I want you to ask yourself is this loan going to get me closer to financial independence and if it's not, ask yourself, is this the right time for something like this because you will be forced to work to pay bills instead of working because you want to prosper and enjoy those elements of life that can only be experienced with a low stress life.

One of the biggest mistakes I see is when a service member retires is to commit future funds for today's toys. People just can't wait…adults wait. They should secure their future first, then consider limiting the number of expensive toys they get rather than as many as they can get.

In other words, one toy at a time. While it is true that toys may be able to be sold if needed, my response is that it's not that simple once you've started accumulating toys and you're used to them because many people change their standard of living and don't want to take a downgrade.

So, as a result, they will continue to work and always stay in this financial home with their expensive toys. The biggest perpetrators of these are the people who go overseas and get that $250,000 a year job. These people get these high paying jobs and continue to accumulate bills and toys but don't have the time to use them. They are working four months overseas one month home, four months overseas one month home.

The problem with that is that many of those people are not getting paid when home, and they create a financial requirement of having to go back overseas to pay for the lifestyle. In many instances their significant others are enjoying these toys and homes without their spouses because they're not here to enjoy them together.

In other words, their significant other is taking the boat out by themselves because he or she is overseas earning the money to pay for the boat, so be careful once you get that new job in creating new bills because you also may find yourself committed to have to work many more years in order to maintain that lifestyle.

I think the real warning label here is that the overseas job is not guaranteed and may go away. People should be financially focused on paying off their bills that they currently have, in the event that they can't go back overseas. Furthermore, what if the member comes home and is disqualified from going back?

I don't want to be a dooms day person but what about car accidents? What if your children needed attention, such for unexpected medical changes? Let's take a look at Frank.

Frank B., US Navy, 4 years of service

"When I got out of the Navy, I was blessed enough to have 50% disability which at the time is it about $400 a month. I was able to use my military benefits and get a terrific job at a local airbase that provided me more than enough money to live a very comfortable lifestyle.

The problem was I got a little too excited and bought a big piece of land that I could barely afford while building a home on it to support my family. Since I can get fired at any time, the only guaranteed money was the veteran's administration $500 a month.

Instead of working with the money I had from my job incrementally to satisfy my future financial needs, I decided to finance the home build which put me in financial shackles for the next 25 years. At least that's what I thought... I experienced a divorce. In the divorce, I did get the house as part of the divorce settlement, but I had to pay cash to my ex-wife which force me to get another loan to pay her equity.

This increased my house payment because I refinanced the house to pay for this bill which means another five years. If I had financial discipline, I would have bought a home that made more sense that I could live within my means and focus on what's really important because in the end I didn't get to keep my wife, but I do have home on a piece of land that is always going to need some sort of repair."

As you can see by Frank's example, not taking on

more financial responsibility than you need to would've made a lot more sense for him given the circumstances. I know many other people like Frank that had more positive examples but that is the luck of the draw for some.

Divorce and splitting your assets in half is quite common in today's world, and things don't always work out the way you wanted to, so not putting yourself in a financial situation may be a good solution for you. Don't bite off more than you can chew.

In closing in his financial responsibility section, keep your bills low, don't go in car new debt before you separate are retired, and don't take on more than you can chew.

Once you get that job after you separate give yourself a little time to ensure you know how much money you're going to have and make financial decisions that are going to carry you forward to the next phase of your life.

After all, shouldn't we be removing stress once we leave the military rather than adding stress? Removing financial debt is a terrific way of removing stress from your life

Dealing With The Veteran's Administration

How to initiate and work with the VA (veterans affairs) that is essentially an unknown to most while starting their transition. I will talk through points and share my own and other's experiences to deal with this intimidating task. This being a huge part of you taking your next step towards your civilian transition. You have done so much for this country already, dealt with injuries and pains that are a result of your service so, why not be compensated for them.

Most individuals don't take advantage of this entitlement. I don't know if it is a pride issue or not wanting to "take advantage of the system." Whatever the reason it is not fair to you as a service member who has sacrificed so much to this country. If you are not getting your aches, pains and past injuries documented in your records you are wrong.

Here is an experience from a prior service member that had an issue dealing with the VA.

Kayla C., US Air Force, 4 years of service

"Although I only did a few years in military service, it still took a toll on my body. I did not really submit or take advantage of my VA medical benefits right away when I separated.

This proved to be more painful down the road to get issues that arose that were service related the attention required from the VA. The things that seemed minor when I was in, came up down the road and I felt hopeless getting them addressed. Had I only accomplished the small amount of leg work when separating a lot of my issues would have been documented.

The thing I didn't understand was even if it was a rating of zero, it is extremely beneficial to receive treatment for that issue should it arise later down the road. Now I was able to get some of my issues resolved later down the road, it was much more difficult than it needed to be. Not only the paperwork but the time to get the issues looked at.

Another thing that I didn't know at the time, you have up to one year from separation to submit your claim to receive the back pay for the issue. If you go past a year, you only receive pay from when the claim was initiated. That would have been helpful at the time to know.

I am sure someone told me this or it was in some paperwork, with the hustle and bustle of separating a lot of things I wish I would have paid more attention to fell of my radar."

Kayla had an all too familiar experience with the VA, transitioning members often become overwhelmed with separation process. Information is passed so rapidly and, in such bulk, it seems almost impossible to retain it all.

I will try my best to make the information that much more palatable with a few steps and hints that helped me with the process and what to look out for.

First step is submitting an intent to file with VA.gov. this can be accomplished by going on to the site and fill out required documentation. A Veteran service center (VSO) can help you with this step as well. Next is gathering the necessary medical documentation. How I accomplished this was went to the medical records management office and asked for a copy of my medical records. Once I had my records in hand I went home and went page by page, line by line and put sticky notes by each and every thing I wanted to highlight.

Next, I made an excel spreadsheet with all my bodies aliments that were in my records and recorded what was currently hurting. After I was done with that tedious task, I walked into the doctor's office to address them all.

At the appointment after the doctor asked me what brought me in today, I handed her my spreadsheet, and her response was Maj Martinez you must be retiring. I said, well as a matter of fact I am, she looked at me and said OK, well I will address all the issues on this spreadsheet, and I will refer you to the proper specialist to take care of all of this. To my surprise that is exactly what happened, and I was able to get everything documented in my records. My first step into my VA journey was completed.

Dennis had an alternative experience with the VA and let's see what happened with his claim.

Dennis H., US Air Force retired, 20+ years

"As a retired medical technician, I saw the worst medical

cases people can expect to see day in and day out. I took care of people who truly needed medical attention.

When I retired, I didn't see myself in the same situation as the people I cared for, so I never pursued my medical conditions because of who I treated so I was happy with my initial claim of 40% despite the fact I thought I was more disabled that VA stated.

Since I used a VSO, I took the opinion of him when he said he would wait to request more VA at a later date. Unfortunately, I waited 6 years before claiming my actual benefit that eventually came in at 100%.

This delay in my submission cost me $183,000 in lost cash claims not to mention both of my sons would have gone to college for free which adds up to another $30,000 that we *paid for tuition…all this because I did not pursue my total claim immediately!*

My recommendation is to stay on top of your VA claim because no one cares more about your claim or future more than you. It is not the VSO's job to redo your claim, or follow up on issues, it's yours! Stay on top of it because your future depends on it."

As you can see from Dennis's experience do not downplay those issues, even if you do not think it, is an issue. Don't let anyone influence you on your entitlements and do what you need to do for your medical needs down the road.

Depending on the state you live in one-hundred

percent disability will pay an amount for your children's college up to 36-45 months of fulltime enrollment, as well as not having to pay property taxes depending on

Now for the third step, talking to the VA and starting that process. There are plenty of resources to help you with this process. Again, I would strongly suggest reaching out to a Veteran Service Officer to help guide you through the process. There are plenty of options on base or off base just ensure the VSO is accredited.

It is important to not to have a macho attitude when discussing your issues, don't lie about anything but don't underplay any of your issues either. By downplaying your issues will hurt not only your future medical representation but also your deserved compensation.

For example, when I was asked about my back the nurse asked me to bend over and touch my toes but stop when it hurts. Now sure I can bend down and touch my toes but that is not what was asked. I bent over and barely moved 6 inches and I stopped. The nurse looked at me and said, "you can't touch your toes?" I said, "sure I can but that is not what you asked, you told me to bend over till I felt pain, this is how far I can go before I feel pain."

Make sure you are understanding and speaking the language of the physician when being asked. A great piece of advice I can give on this subject is to talk with individuals who are currently going through the process and those who already have. Hopefully this will help prepare you for what is to come better.

Lastly make sure you are logging in and checking on the claim. Sometimes the VA during their investigation process will require more documentation or signatures. Ensure that you are staying on top of it and checking it frequently. It is not just a submit and forget it situation.

Let's see what my good friend Chet had to say about his experience with the VA and what his opinion on the issue is.

Chet P., US Navy, 2 years of service

"As a cook for the Navy, I discovered that the sailor life was not for me. I served a two-year commitment and immediately separated honorably as fast as I could.

As the years went by, I never considered that such a short time in the military could provide such benefits such as VA medical payments.

About 50 years after separating, I was talking to my nephew who encouraged me to get my Navy medical records to see if I qualify for any medical benefits... turns out I qualified for 10% medical disability.

The payment on 10% is only $100 or so per month but remember, I didn't cash in for over 50 years which means I missed out on 600 months of payments which means I did not get and, I am not entitled to the back payment of $60,000!

My recommendation to all separating members is to submit for your VA benefits when separating and have the VA tell you that you don't qualify rather than assuming you don't qualify and miss out on your entitlements. Many of my friends told me to do it but I didn't see the point, but in hindsight I wish I would have filed for my benefits the day I separated."

Chet is just yet another example of not correctly utilizing the benefits that are given to you for your service to this county. $60,000 is a lot of money that could have been used to better his life and lifestyle after his transition.

Dealing with the VA isn't exciting or easy by any stretch of the imagination. However, hopefully if you utilize these steps and tip's, it could make it just a bit less daunting. In closing remember to speak the language, do not downplay anything no need to be prideful with this and do not procrastinate your claims.

Job Searching

In my opinion this topic is the cause of most of our anxiety and fear when transitioning. The question of what do I want to do? The unknown feeling of what the job market is doing currently. The anxiety of understanding what can I bring to the table for a company to hire me. Will I be able to find a job.

Having faith in yourself and putting yourself out there is the first step. You have a wealth of experience and leadership qualities that you have learned over the course of years in the military. Let that calm your nerves for a minute. I made a pro and cons list of a few jobs that I was interested in pursuing, this allowed me to not only think through options, but I now had a visual to help make my decision off of.

Step two, get that resume drafted up in a template format. This is important to be able to mold the resume for the employer and job you are targeting. Most of the time today with advancement in technology resumes are ran through a computer program that use algorithms to narrow down probable hires.

This algorithm focuses on keywords to allow the company to find someone that meets the criteria they are looking for. With that being said you need to ensure your resume is targeting the keywords from the job's description/advertisement.

For instance, if you are looking for a job as

cook and the restaurant is a steakhouse and is looking for someone that has experience grilling steaks, your resume better say something about your expertise in grilling steaks.

I know that is a basic example but make sure your resume is targeting the needs and expectations of the company. Also make sure if you're telling a company that you can grill a steak you better know how to grill a great steak.

Now that you have a resume, utilize it and put it out there. A common mistake is not applying to multiple companies, now when I say multiple, I mean as many as you can. Do not just apply to one organization and think that could be it. Is there a chance they could hire you? Maybe, but that isn't a chance I would take. I want options, options that I can weigh the pros and cons of. I want the interviews to understand the company and its goals and what my role would be in that company.

I am not saying to be so picky and turning down jobs and not taking opportunities as they arise. Just like companies can fire you, you have the choice to fire them as well. Taking a job outside of the military does not mean you have to stay in the company for the next 20 years, you have the ability to move around take different opportunities. You are not locked in by contracts necessarily like in the military.

You may have to take a job you don't like and that is alright, who knows maybe after some time you may learn you love it. All I am saying is that if you are in a job and you don't really like it, then just keep looking for another job and if one arises put your notice in and move on to greater things.

The importance of staying fluid and keeping options open and an open mind are extremely important with your transition. Now here is a take on the importance of the job search and being able to keep an open mind from Josh.

Joshua A., US Airforce retired, 20+ years of service

"Dig out all your old EPR's or OPR's. Use your job descriptions and other bullets to expand your job experience on your resume. Just saying you spent five, ten, fifteen years as an aircraft mechanic won't take up but one or two lines.

Make sure you include additional duties you had.

Love them or hate them, they were a valuable experience in leadership, administration, and other technical skills. They made you a better writer, computer user, mentor, and manager. Make sure potential employers know you have the experience.

Cast a wide net. When I submitted my retirement paperwork in August of 2019, I knew I was a shoo-in for any pilot job I wanted.

In December and January, I started collecting letters of recommendation and filled out applications to four airlines. Just four. Three that I really wanted to work for and a fourth as a back-up because it required zero effort on my part.

Well, we all know what happened in February and March of 2020. COVID. Every major airline stopped hiring and training, including all four that I applied for.

So, I had to start over. I applied to 18 more companies.

Including several that I had no real interest in working for, but they would provide a paycheck until the world got back to normal in a few months or years.

Fortunately for me I found a company that was hiring and training. You never know what the world is going to throw at you, so be prepared. Had I cast that wider net to begin with, I would have saved myself a lot of stress and scrambling to fill out more applications at the last minute. Don't be like me, be ready for anything.

Be open to other opportunities. Like I said, I thought I was a shoo-in for any pilot job I wanted. I would have been, if COVID never happened. The timing of my retirement forced me into a job that I hadn't considered.

I didn't really have any other choice at the time. As it has turned out, its pretty much my dream job. It's easy and it pays well and offers opportunities I would not have received at the four jobs I first applied to. There are downsides, yes. But know two years into it, I have no intention of changing jobs.

Approaching your retirement or separation, you will hear a lot of good things about a lot of your friends' jobs. Listen, take what they have to say and take it into serious consideration.

Ask what they don't like about their job or company. They may be reluctant to tell you… but ask. Every job has them. Do the research on your own. Look at online reviews. Look at corporate portfolios, this can give you an understanding of how the company has performed financially over the long term.

Whatever qualities about a company that matter to you. Find the answers to it. Just weigh the pros and cons that you know and make the most informed decision you can.

More importantly keep an open mind. Going back in my situation, if you cast a wide net of applications and are ready for whatever happens you'll be better prepared. You never know when a job you would have never considered in the years or months leading up to getting out will turn out to be your dream job."

As you can see from what Josh shared, unexpected things do happen, and he changed up his approach and was able to get a great job that he loves by as he says, "Casting a wider net". Nothing is guaranteed and be ready and prepared to roll with the changes and adapt to whatever is thrown at you.

Being able to not close your mind off to opportunities as they arise is another trait, I would like you to entertain. Moving out of our comfort zone is a tough and scary thought but may just be the key that opens many doors for you. These doors prior could have been locked prior and that opportunity would have never available without that one key.

Back to keeping your options open, try it, explore the unknown and you never know what could be awaiting you. This can totally back-fire and you could absolutely hate it. Good news is that it doesn't have to be permanent but at least now you know, and I am sure you now not only have made contacts in that field, but you have knowledge and experience in it as well.

A great example of that is from Victoria's experience of stepping out of her comfort zone. Let's read about her experience and how it was helpful for her.

Victoria G., US Air Force retired, 20 years of service

"Professionally I was an enlisted Air Force Aircraft Maintainer: I'd have to say one of the biggest things I that was beneficial for me was to learn my strengths while I was in service, what was I good at and what did I enjoy.

Those things turned out to be leading people and managing projects. The second thing was humility. These two things tie together once it come to transitioning even more so the higher rank you held.

The joke is always "what do you want to be when you grow up" aka what are you doing when you get out of the military. I chose Project Management and let me tell you I was extremely blessed with my first job when I retired. I was contacted by a recruiting firm to fill a contracted position with a large company in the Mobilization and Transition department.

I had no clue who or what this company was, let alone what mobilization and transition meant to them. In the military it military it's moving from one place to another. Turned out it's a similar concept in the industry I was in, instead of mobilizing people and aircraft I was mobilizing information from one software program to another.

Starting this job was not easy. I was so nervous with no clue what I was doing, what I was going to wear and hoping I can adjust to the culture. I started working mid-tier and I immediately became humbled. I was a "go to" for so long on how to get help for this computer issue, that company policy and how to work 100% remotely. Funny story: if you take a contracting job, you must log your hours like a timecard.

Last time I did that was on a card that you would slide into a thing, and it would stamp the date and time for you! Not on this.... I had codes I was supposed to choose from that didn't exist because they were not loaded. Five zoom calls later we found someone that knew what to do and was able to help me.

Welcome to week one on the job and being humbled... I had lots of rapid-fire humbling moments after that, things as simple as "what are you even saying" they were speaking in accounting and software development terms. This team took a chance on me!

I had no clue about software or accounting, but I knew how to run a project and after I got the terminology down and learned the KPMG and teams' culture, I started gaining traction. Now, I absolutely love that I did something that I have never experienced before.

I was humbled and forced to remember that we all must start somewhere. It's ok to be an expert at somethings and stumble over others. The team I worked with were patient and understanding, they guided me expertly through the learning curve and taught me what was pertinent.

One thing that I will ask is if a vet reaches out for an informal interview, please respond to them. If you don't have time, connect them with someone who does or a resource that can help get them connected.

The transition process is filled with uncertainty sometimes just having someone giving them an example of what it's like in the civilian sector can ease a lot of stress. Finally, Take some time for yourself.... Decompress and reflect this is a new chapter of your life that you have full autonomy over, do it your way!"

Victoria's recommendations are spot on with humility in searching for that job. Her ability to step out of her comfort zone proved beneficial for her next career. Too many transitioning members are arrogant or too shy...the secret is the balance between the two to make a confident impression on the hiring official. Let's take a look at another suggestion.

Hank H., US Air Force retired, 20 + years of service

"When I retired I thought I wanted to get an easy office job and work to make my bills. I was fortunate enough to land a job as the City Manager for my county. The job provided many of the same feelings I had while I was in the military, service, a sense of importance, and security. After only 11 months I realized that what I really wanted was freedom, so I quit. That's right, I had a warm, cushy job and

decided to pursue my own endeavor as a YouTuber.

I started my channel small and doing things I enjoy, like farming and working around my property and now I'm a very well paid YouTube personality with over 274,000 subscriber making way more money that I ever dreamed and doing it all on my time! Exactly what I was looking for. My suggestion to you is to make sure you follow your heart and use your retirement and any associated VA benefits to provide you the insurance you need to allow you to take a calculated risk— Your Future Happiness May Depend On it!"

What an inspiring outlook on Hank's next chapter of his life. He did not allow his ego or tunnel vision to pigeonhole him into a 9-5 job punching in and out of a boring routine life. He ceased his moment and with the security of his pensions he is following his passion and he now can have his cake and eat it too.

Throughout this last chapter we discussed the severity of the dreaded job search. For most of us we haven't had to do this in at least twenty plus years, making it that much more intimidating.

In summary to our discussion. Get that resume tight and accurate targeting the job you are applying for. Apply to multiple (as many as you can) potential employers "cast a wide net". Know that the position doesn't have to be permanent, don't throw away opportunities, you can always be on the hunt for the next best thing.

In Closing………..No one cares more about your future than you! You need to be proactive and plan your future with a deliberate and informed set of facts rather than trying to figure things out assumptions as you go along. Please take the time to expand your knowledge on these areas and create a plan to ensure you and your family receive all the benefits you are entitled to.

We recognize this book just covers a few issues service members should know and there are more… many, many more. Your stories and advice should not be quarantined in your brain, please share them with us at info@YourInstructor.com so we can continue to update these experiences for future transitioning members.

We hope this book opened your eyes and made you future brighter.

www.ingramcontent.com/pod-product-compliance
Lightning Source LLC
Chambersburg PA
CBHW070612010526
44118CB00012B/1494